Math Concept Reader

Miss B.'s Class Makes Tables and Graphs

Developed for Harcourt, Inc., by Gareth Stevens, Inc.
This edition published by Harcourt, Inc., by agreement with Gareth Stevens, Inc.

Printed in China

ISBN 13: 978-0-15-360222-1
ISBN 10: 0-15-360222-8

12 13 14 15 16 0940 20 19 18 17 16
4500590863

Miss B.'s Class Makes Tables and Graphs

by Joan Freese

Harcourt
SCHOOL PUBLISHERS

It is Be Healthy Month. Children learn how to be healthy.

They learn about exercise. They learn about foods. The right foods help them grow.

Miss B.'s class gathers information.
Information is also called data.

The data will answer a question. Which fruit does the class like best?

Miss B. draws a tally table. One row is for apples. Another row is for grapes. Another is for oranges.

"Which fruit do you like best?" Miss B. asks. Each answer gets a tally mark.

The tally table is done. The class counts the tally marks. The children like grapes best.

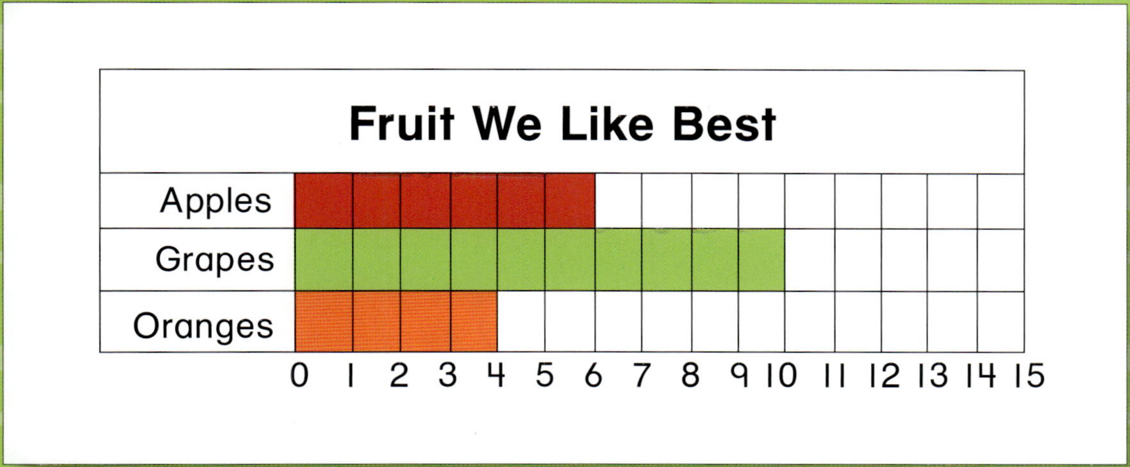

Next they make a bar graph. They use data from the tally table. The bar for grapes is longest.

Miss B. asks, "Which drink do you like best? We can gather the data."

The class makes another tally table.
Each child picks milk, juice, or water.
They count the marks.

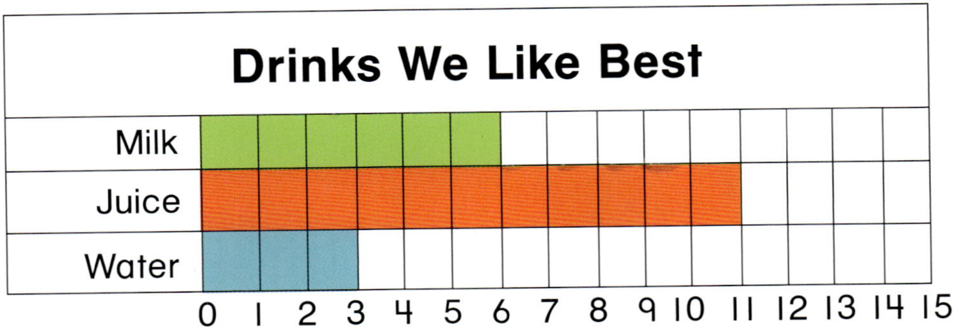

They draw a bar graph. How many people picked milk? Six!

How many people like lemonade? The class is not sure. “The graph does not tell us,” Miss B. says.

Miss B. asks her class to think. “It is Be Healthy Month. What else can we graph?”

Sam says, “We can graph the exercise
we like best.”
Miss B. says, “Good idea!”

The children know what to do. They make a tally table. Then they mark the exercise they like.

A bar graph comes next. What do they like best? Biking! What do they like least? Walking!

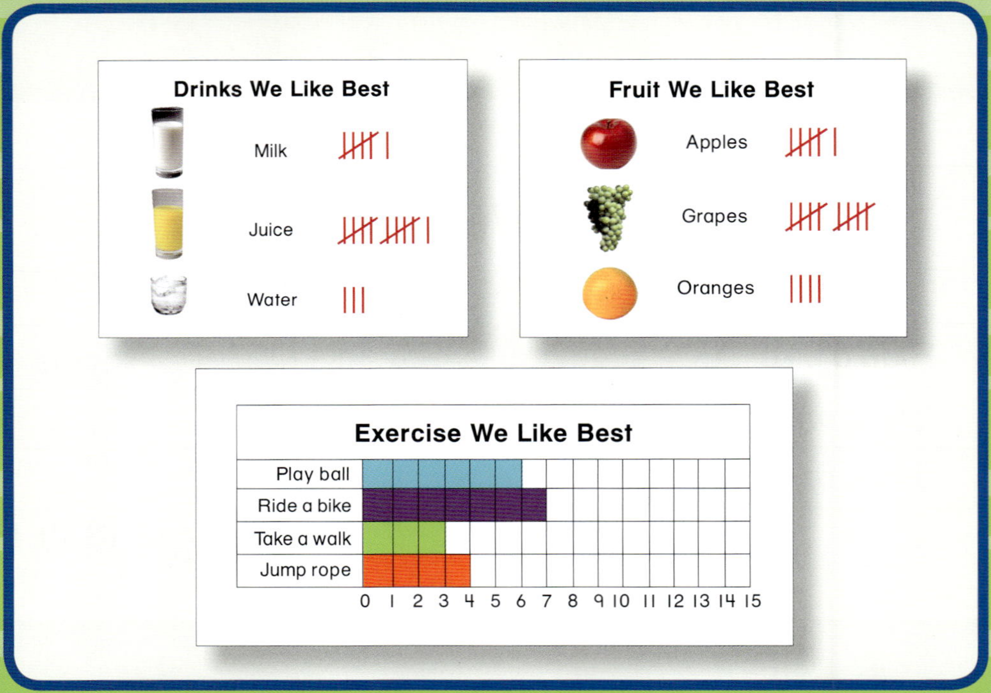

Miss B. is proud. Her class can make bar graphs. She hangs their work in the hall.

Other classes read the bar graphs. They learn what Miss B.'s class likes best.

Be Healthy Month is over. Miss B. has a party. She brings snacks.

The class heads outside. “Playtime!” they shout. Miss B. smiles.

Glossary

data information collected about people or things

exercise activity that helps to keep the body healthy

information facts or data about a person, place, event, or thing

Photo credits: cover, title page, pp. 3, 5 (center), 10 (left); U.S. Department of Agriculture; pp. 2, 13 © Felicia Martinez/Photo Edit; pp. 4, 14, 20, 21 © Michael Newman/Photo Edit; pp. 5 (left), 10 (center), 19 (right) Gregg Andersen; pp. 5 (right), 6 (all), 7 (all), 8 (all), 11 (all), 15 (all), 18 (all), 19 (left) © Hermera Technologies, Inc.; pp. 10 (right), 19 (center) Joe Major/DK Stock/ Getty Images.